High Interest/Low Readability
Biographies

Frank Schaffer Publications®

Author: Delana Heidrich
Editors: Linda Triemstra, Raymond Wiersma

Frank Schaffer Publications®

Send all inquiries to:
Frank Schaffer Publications
8720 Orion Place
Columbus, OH 43240-2111

High Interest/Low Readability Biographies

ISBN: 0-7696-3394-3

3 4 5 6 7 8 9 10 MAZ 10 09 08 07

Table of Contents

Introduction

Middle school students love a juicy story. A true tale of triumph over poverty, misfortune, bad choices, or failures makes a juicy story. Learning that Abraham Lincoln was a lousy postal clerk before becoming one of our most beloved presidents feels more like listening to gossip than reading history. Uncovering the rags-to-riches life of Hans Christian Andersen resembles reading a fairy tale more than plodding through a biography.

High Interest/Low Readability Biographies contains juicy true tales about the bumps along life's road as encountered by men and women. Students will recognize these individuals by their eventual successes in literature, politics, business, sports, or the arts.

Each story is preceded by prereading activities and followed by comprehension and follow-up activities. Use these complete lessons to motivate and encourage your low readers while providing them practice at the skills needed to read for meaning, insight, and enjoyment.

A Life of Adventures: Miguel de Cervantes Saavedra • Rags to Riches: Hans Christian Andersen and Alice Walker • The Detective Who Wouldn't Die: Sir Arthur Conan Doyle

Mr. Chocolate Kiss and Colonel Chicken: Milton Hershey and Colonel Sanders • Babes at Bat: Babe Ruth and Babe Didrikson • The Greatest Showman on Earth: P. T. Barnum

American History Unearthed: Abraham Lincoln and Thomas Paine • Old Ladies on the Move: Maggie Kuhn and Mother Jones

Prereading Activities

Looking It Over

1. Read the title of the article that begins on page 6.

2. Leaf through the pages of the article, stopping to look at the pictures.

3. Read the list of vocabulary words.

What Do You Know?

Record here all you know about Miguel de Cervantes Saavedra or *Don Quixote;* Spain in the 1500s; pirates; or knights.

_____

_____

_____

_____

_____

Make a Prediction

What kind of adventures might Miguel de Cervantes Saavedra experience in this story?

_____

_____

_____

_____

_____

Vocabulary

scrounging   seeking out and taking

Example: Mom is scrounging around her purse for enough change to buy a soda.

malarial   caused by malaria, a disease

Example: Andy was kept in bed for a month with a malarial fever.

ransom   a payment made for the release of a captive

Example: The kidnapers will release Anna if we pay the ransom.

armada   a fleet of ships

Example: The Spanish Armada conquered the seas.

antics   foolish and playful acts

Example: We laughed at the clown's antics.

scholars   educated people

Example: History scholars know the names and dates of the important events in their nation's history.

A Life of Adventures: Miguel de Cervantes Saavedra

Miguel de Cervantes Saavedra was born in Spain in 1547. He lived in exciting times. Spain was a great power in Europe then. They had just found the New World. They had conquered most of South America by 1550. Now they were moving north into Florida and Cuba.

During Cervantes' adult years, Spain focused on closer conquests. They formed the Holy League. This army fought in the name of Spain and God. The church and the state were strong when they acted as one. They fought France and Italy. With a 130-ship fleet, they even tried to take over England.

Cervantes got caught up in his times. He lived a life full of adventures. Then he wrote about them. Spain's best-known author did not begin telling stories until the age of 50. He was too busy before that!

Cervantes' adventures started early. He was born into a family of eight. His father was a doctor who went from town to town treating patients. Some of them could not pay. So young Cervantes and his siblings spent more time traveling and scrounging for food than going to school.

As a young man, Cervantes joined the navy. There he didn't have to scrounge for food or clothes. But Cervantes' rough times had just begun. Now he found himself on a battleship. He was going to fight for the Holy League. They were about to face the fierce Turks. Before the battle began, Cervantes became sick. Now he had to fight both his enemy and malarial fever. The fever made Cervantes very sick. Still, he fought hard in the Battle of Lepanto. In the end, 100 Turkish ships were captured. Thousands of Christian slaves were freed. In the battle, 25,000 Turks and 8,000 Christians were killed.

Cervantes lived through the battle. But enemy fire did catch up with him on the ship's deck. Two bullets struck him in the chest. Another shot through his left arm. His left hand and fingers were useless for the rest of his life.

7 0-7696-3394-3 *Biographies*

At last, Cervantes' ship began its return trip to Spain. On the way, pirates attacked it. Most soldiers were sold into slavery. Cervantes was held captive by the pirates. He had a letter in his pocket. It was written by one of his bosses in the navy. It said Cervantes was a brave soldier. The letter was supposed to help Cervantes get a job when he got home. Instead, it made the pirates believe Cervantes was an important man. They asked his family for a high ransom. Even with the help of the church, it took them five years to raise it.

Back home, Cervantes decided to write. One short story about farm life and a few plays earned him a little money. But he needed more than a little money. He wanted to pay the church back for the ransom it helped raise. So he went to work for the crown.

He took wheat and olive oil from farmers for use by the Spanish Armada. Farmers did not like parting with their crops. They mocked and threatened Cervantes. People in town were not nice to him either. He was even kicked out of his church. So Cervantes asked permission to go to the New World. He was turned down. Instead, he became a tax collector. Without much schooling, he did not work well with numbers. Twice he was put in jail for sloppy bookkeeping.

Jail was Cervantes' lucky break. During his second stay, he turned to writing again. This time, he wrote of what he knew—travel, war, ships, and prison. His fictional tale told of the antics of Don Quixote. Quixote thought he was a knight. Many funny things happened to him. Readers liked Cervantes' humor. They bought his book in great numbers. Cervantes wrote more tales. Still, Cervantes made little money from his works. The books people bought were often illegal copies that made money for other people.

Cervantes died a poor man. Still, the tales he wrote had lasting power. *Don Quixote* is known as the world's first true novel. It is a classic. Scholars talk about it. Translators write it in many languages. The wild early life of Cervantes may be long forgotten. But the adventures of Don Quixote live on.

Comprehension

Circle the best answer. Highlight the sentence or sentences in the story where you find each answer.

1. Miguel de Cervantes Saavedra was born . . .
   a. into a small family.
   b. into a rich family.
   c. in Spain.
   d. in 1847.

2. Cervantes' father was a . . .
   a. doctor.
   b. lawyer.
   c. shoemaker.
   d. author.

3. While suffering malarial fever . . .
   a. Cervantes was sent home from the navy.
   b. Cervantes was put in jail.
   c. Cervantes wrote *Don Quixote.*
   d. Cervantes was shot three times on the deck of a battleship.

4. Pirates held Cervantes for ransom because . . .
   a. his family was very rich.
   b. he held a letter that made the pirates believe he was important.
   c. they held all the men on Cervantes' ship captive.
   d. Cervantes was a very important man in Spain.

5. When his ransom was finally paid, Cervantes . . .
   a. wrote a book about his adventures.
   b. worked for the government.
   c. became a tax collector.
   d. went to America.

6. While working for the crown, Cervantes . . .

    a. collected crops for the Spanish Armada.

    b. was well liked by the farmers in Spain.

    c. visited the New World.

    d. acted as the queen's servant.

7. Cervantes went to jail because . . .

    a. he didn't pay his bills.

    b. he stole money from a tax collector.

    c. he didn't pay his taxes.

    d. he didn't keep good records as a tax collector.

8. While in jail, Cervantes . . .

    a. had many adventures.

    b. got in a sword fight with his cell mate, Don Quixote.

    c. wrote a book.

    d. painted his first masterpiece.

9. Cervantes died . . .

    a. a poor man because he got no money from the sale of illegal copies of his book.

    b. a poor man because none of his books sold until after his death.

    c. while in jail.

    d. in a battle as a naval officer of the Spanish Armada.

10. *Don Quixote* . . .

    a. is no longer in print.

    b. is considered the world's first epic poem.

    c. is a classic novel that continues to be read today.

    d. is the true story of a stern and brave knight.

A Life of Adventures: Miguel de Cervantes Saavedra

Discussion Questions

1. Cervantes lived a very full life. Which of his adventures would you have enjoyed the most? Which adventures would you not have liked?

2. Cervantes first started writing books while in jail. Tell about a time you turned a bad experience into a good one.

3. Cervantes did not have the use of his left hand for most of his adult life. Tell about a person you know, have read about, or have seen in a movie who has accomplished great things in spite of a physical challenge.

Sequencing Scenes

Place a number in each of the spaces below to indicate the order of Cervantes' adventures.

_____ Cervantes writes *Don Quixote.*

_____ Cervantes is shot in the chest and arm.

_____ Cervantes travels with his father, who treats patients in many towns.

_____ Cervantes is mocked and threatened.

_____ Pirates hold Cervantes captive.

_____ Cervantes is a tax collector.

Vocabulary Activities

Complete each sentence with a word from the Word Bank.

> **Word Bank**
>
> scrounge   ransom   malarial
>
> armada   scholars   antics

1. Cervantes' _____ fever made him very sick while on board a battleship.

2. Don Quixote's _____ continue to capture the attention of readers today.

3. If I _____ around my closet, I might be able to find the lost slipper.

4. The _____ note stated Tiffany would have to pay $25.00 to get her dog back from the neighborhood bullies.

5. Literary _____ call the original Don Quixote story the world's first true novel because its well-developed plots and characters were meant to entertain, not teach a lesson.

6. Kelly had an entire_____ of toy ships in her bath water.

Homework

Cervantes lived with pirates for three years. Write a letter from Cervantes to his parents telling them about his time spent in a pirate prison.

Extension

Locate a Don Quixote tale in your school library to read aloud in class.

Prereading Activities

Looking It Over

1. Read the title of the article that begins on page 16.

2. Leaf through the pages of the article, stopping to look at the pictures.

3. Read the list of vocabulary words.

What Do You Know?

1. What do you know about Hans Christian Andersen? List the titles of any Andersen fairy tales you know.

_____

_____

2. What do you know about Alice Walker? Can you name the title of her most famous work?

_____

_____

3. What do you know about the civil rights movement? What was life like for blacks who lived in the South before the Civil Rights Act?

_____

_____

Make a Prediction

What might the lives of Hans Christian Andersen and Alice Walker have in common? Why might they be famous?

_____

_____

Vocabulary

poorhouse   in earlier times, a home where poor people lived

Example: People with no jobs in nineteenth-century England lived in the poorhouse.

recited   repeated from memory

Example: Amanda recited the poem in class.

trade   a job requiring certain skills

Example: You will always have work if you learn a trade.

racial slurs   insulting remarks based on color or culture

Example: The only Polish family on the block suffered many racial slurs.

sharecropper   a farmer who works on another's land

Example: Sharecroppers do not make as much money as do landowners.

plight   bad situation

Example: Environmentalists are concerned about the plight of endangered species.

inspired   encouraged and enlightened

Example: Deb's kind actions toward others inspired me to be kind as well.

Rags to Riches: Hans Christian Andersen and Alice Walker

Hans Christian Andersen and Alice Walker are very different authors. They lived at different times. They grew up in different countries. They did not write the same kinds of stories. But they share one thing. They both turned a rough start in life into success.

Hans Christian Andersen

Hans Christian Andersen wrote fairy tales. He also seemed to live one. Andersen was born in a one-room house in Denmark. The year was 1805. His father died when Andersen was 11. He left no money behind. He did leave his son books, puppets, a toy theater set, and a love of tales.

Andersen liked to hear and tell stories. Old ladies in his town's poorhouse told tales while they spun yarn. Andersen recited their tales by heart. But he could hardly read or write. So he did poorly in school. Instead, he chose to learn a trade.

Andersen failed at one trade after another. He was not a good weaver. He was not a good cigar maker. He was not a good tailor. He was not a good shoemaker. He was already a four-time failure by the age of 14. So he packed his bags and moved to the city. He wanted to sing, dance, act, and write plays. So he did. He was not good at acting or dancing. He could sing for a few months, until his voice changed. Then he wasn't good at that either. And no one wanted the plays he wrote. They were too full of spelling and grammar mistakes.

For three years, Andersen relied on friends to feed and house him. Then the director of a local theater saw through the mistakes in one of his plays. He could tell Andersen knew how to write a great story. He just needed to sharpen his skills in spelling and grammar. The man arranged for the 17-year-old Andersen to attend school.

For six years, Andersen worked beside kids half his age. He learned to write without so many mistakes. At the age of 23, he passed a test to get into college. One year later, he was publishing stories. Five years later, the king of Denmark was paying Andersen to tell his stories all over Europe. The poor boy who couldn't read or work at a trade was now a rich and respected author.

Andersen died at the age of 70. By then he had written 168 fairy tales. He had journeyed outside of Denmark 30 times. He had made friends with famous and common folks around the globe. His stories had been written in over 100 languages. Andersen lived a fairy-tale life, and he wrote some pretty good ones too!

Alice Walker

Alice Walker understands rough starts. She was born a black child in Georgia in 1944. As a black then, she had to sit in the back of buses. She could not eat in "whites only" cafés. She had to endure racial slurs. Born the eighth child of a sharecropper, Walker suffered poverty, too. She lived in a shack with a leaky roof.

At the age of eight, Walker was playing with her brother when a BB gun pellet struck her right eye. The accident left Walker blind in that eye. It also left her with a scar. The scar embarrassed her. Walker cried when kids made fun of her scar. She did not feel like playing with any of them. So she read and wrote alone in her room.

When she got older, new laws let Walker sit anywhere on a bus. She could now eat in any café. Surgery got rid of the scar on her eye. But Walker never forgot her childhood feelings of hurt and shame. She wanted to make sure others were respected. So she marched with Martin Luther King Jr. She worked in welfare in New York City. She wrote books about the plight of black women.

Walker's novel *The Color Purple* won the 1982 Pulitzer Prize. It was made into a movie. Suddenly the writings of Walker were world famous. Walker had found the best way for her to work for change. Not everyone agreed. Some people thought Walker treated black men too harshly in her books. They said books should not suggest black men do not treat their wives nicely. Walker disagrees. She believes shedding light on injustice will help bring about change.

Walker's work has turned her own life around. It has also inspired others to work for justice.

Comprehension

Circle the best answer. Highlight the sentence or sentences in the story where you find each answer.

1. Hans Christian Andersen and Alice Walker are both . . .
   a. singers.
   b. dancers.
   c. actors.
   d. authors.

2. Which of the following is not true of Hans Christian Andersen and Alice Walker? They . . .
   a. lived during different times.
   b. were both female authors.
   c. grew up in different countries.
   d. wrote different kinds of stories.

3. Alice Walker . . .
   a. marched with Martin Luther King Jr.
   b. wrote fairy tales.
   c. was left blind in both eyes by a childhood accident.
   d. was born in Denmark in 1944.

4. Hans Christian Andersen was good at . . .
   a. weaving fabric.
   b. making cigars.
   c. sewing clothes.
   d. none of the above.

5. Hans Christian Andersen did poorly in school because . . .
   a. he talked too much in class.
   b. he was afraid to speak in front of the class.
   c. he did not read or write well.
   d. he didn't understand math.

0-7696-3394-3 *Biographies*

6. Alice Walker did not like to play with other kids because . . .

    a. they made her feel bad about being poor.

    b. they made fun of the scar on her eye.

    c. they liked her brother better.

    d. they lived too far away.

7. Alice Walker's novel *The Color Purple* did all of the following *except* . . .

    a. discuss the plight of black women.

    b. inspire the making of a movie.

    c. win the Nobel Prize.

    d. turn Walker's life around.

8. As a teenager, Andersen lived with friends because . . .

    a. his mother died.

    b. he was kicked out of his home.

    c. he had no money because no one would buy his plays.

    d. he didn't like being alone.

9. At the height of his success, the king of Denmark paid for Andersen to . . .

    a. write stories about Denmark.

    b. tell stories at the castle.

    c. sing, dance, and act in plays.

    d. tell his stories all over Europe.

10. Alice Walker became blind in one eye . . .

    a. when struck by the pellet of a BB gun.

    b. as an adult.

    c. as the result of an eye disease.

    d. as the result of malnutrition.

Rags to Riches: Hans Christian Andersen and Alice Walker

Discussion Questions

1. Compare and contrast the lives of Hans Christian Andersen and Alice Walker.

2. Andersen went to work at the age of 12. What jobs could you succeed at if you were employed at them right now?

3. How did the times the two authors lived affect their lives?

Diagramming Andersen and Walker

Complete the Venn diagram to indicate what Andersen and Walker had in common and what aspects of their lives were unique to each.

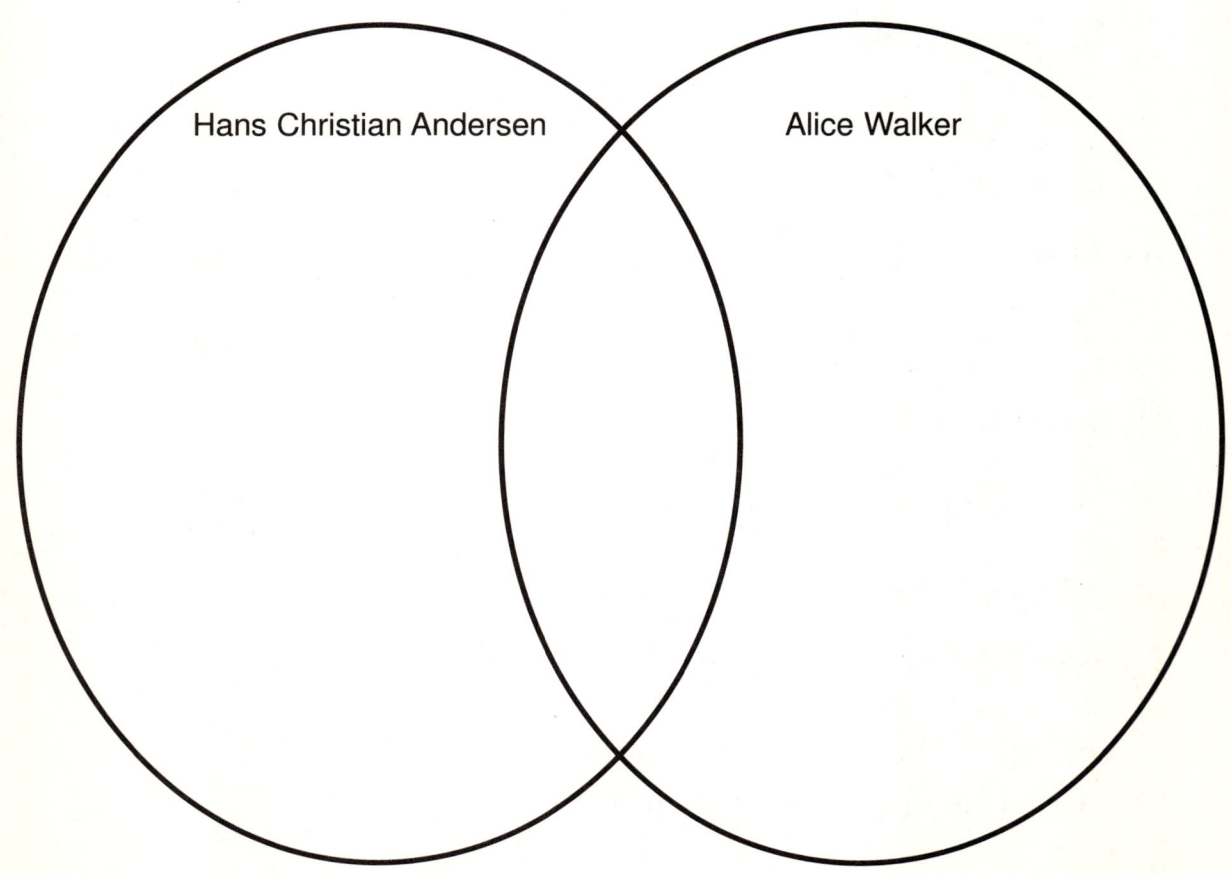

Hans Christian Andersen

Alice Walker

Vocabulary Activities

Write a sentence of your own using each of the vocabulary words listed.

1. poorhouse _____

_____

2. recite _____

_____

3. trade (noun) _____

_____

4. racial slurs _____

_____

5. sharecropper _____

_____

6. plight _____

_____

7. inspire _____

_____

Homework

Write a fairy tale about the life of Hans Christian Andersen.

Extension

Research and report on the differences in fairy tales, folklore, tall tales, fables, historical fiction, and realistic fiction.

Prereading Activities

Looking It Over

1. Read the title of the article that begins on page 26.

2. Leaf through the pages of the article, stopping to look at the pictures.

3. Read the list of vocabulary words.

What Do You Know?

Have you ever heard of Sir Arthur Conan Doyle? Have you ever read a Sherlock Holmes story? What do you know about Sherlock Holmes?

_____

_____

_____

_____

_____

Make a Prediction

What do you think is going to happen in this story about a detective who wouldn't die? Why do you think so?

_____

_____

_____

_____

_____

Vocabulary

featured   given special attention

Example: Kevin watched a movie that featured Eddie Murphy.

fashioned   made in a way to resemble something or someone

Example: Sara fashioned her model car after a real one.

sidekick   partner, assistant, and friend

Example: Saber's sidekick, Roman, goes everywhere he goes.

dissuade   talk out of

Example: I tried to dissuade my teacher from giving me a C- on my report card, but she said that is what I earned.

second-rate   less than the best

Example: The generic brand of my favorite cereal is second-rate.

insisted   demanded

Example: Suzette insisted I sit beside her during lunch.

relented   softened or yielded to a suggestion

Example: When I explained to Mom ice cream would help my sore throat, she relented and let me buy some after all.

The Detective Who Wouldn't Die: Sir Arthur Conan Doyle

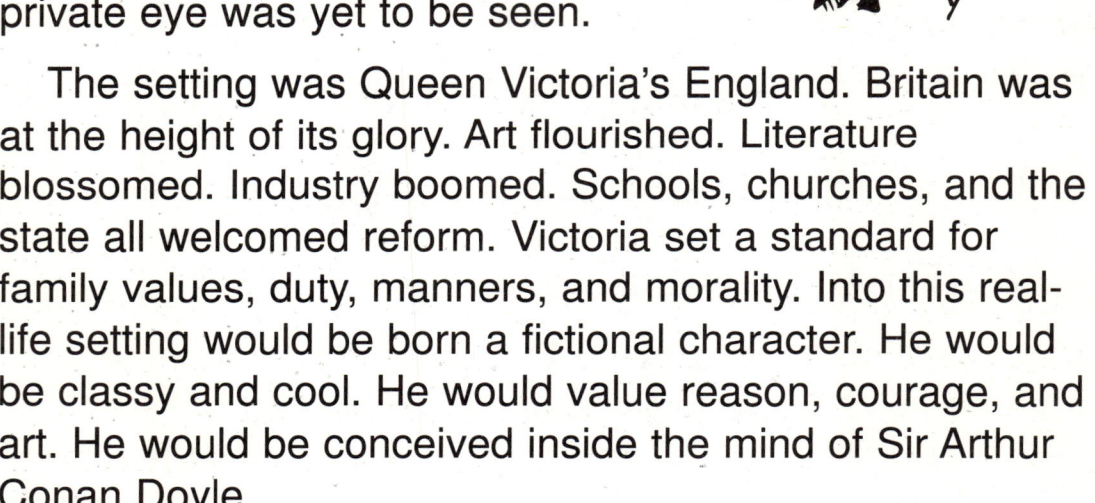

The year was 1838. Edgar Allan Poe wrote a detective novel. It was the world's first. It was called *The Murders in the Rue Morgue.* He wrote more tales in the same genre. In 1868, Wilkie Collins also wrote a mystery tale. In it, a smart detective helped out bumbling police. The mystery story was born. But the most famous private eye was yet to be seen.

The setting was Queen Victoria's England. Britain was at the height of its glory. Art flourished. Literature blossomed. Industry boomed. Schools, churches, and the state all welcomed reform. Victoria set a standard for family values, duty, manners, and morality. Into this real-life setting would be born a fictional character. He would be classy and cool. He would value reason, courage, and art. He would be conceived inside the mind of Sir Arthur Conan Doyle.

From 1876 until 1881, Doyle studied medicine. He wanted to be an eye doctor. By 1882, that's what he was. But people didn't go to the eye doctor often in his day. So his practice did not keep him busy. It did not pay his bills either.

To pass the time and pay the bills, Doyle tried his hand at writing. The short stories he wrote passed the time. But they didn't add much to Doyle's income. That is, until George Newnes read one of them.

"A Study in Scarlet" was a mystery short story. It featured a detective whom Doyle fashioned after one of his college teachers. In fact, he used many ideas from real life to create his story. The detective's name was Sherlock Holmes in honor of the poet Oliver Wendell Holmes. Holmes's sidekick was a doctor with few patients, just like Doyle. Dr. Watson was named after one of Doyle's friends. Holmes's office was located at a London address.

George Newnes read "A Study in Scarlet" in a magazine. He liked it. In fact, he asked Doyle to write more stories about Sherlock Holmes. He wanted one new story each month for one year. Newnes would put the stories in his magazine, *Strand.* Doyle agreed. Writing stories using the same setting and characters should be easy enough. He could do that between patients.

He began the job in 1891. By the next year, Doyle was a rich and famous man. Sherlock Holmes stories were a big hit in England and the United States. The first year's stories were bound into a book. Newnes asked Doyle to write another year's worth. Doyle tried to dissuade him. He loved to write. But he thought the Holmes stories were second-rate. Doyle enjoyed writing medieval romances. He liked to pen science fiction. He wrote political essays. He wrote works on religion. Doyle thought all of these were better than his Holmes stories. So he asked Newnes to pay him 1,000 pounds to write more Holmes stories. He thought he had asked so much Newnes would say no. Newnes said yes.

Doyle kept Holmes alive for another year. Then, in a story called "The Final Problem," Holmes was killed. The public was not happy. Fans begged Doyle to bring Holmes back to life. Even the queen and Doyle's mother insisted Holmes live again. After ten years, Doyle relented. In a new story, he explained that Holmes only appeared to have died. Doyle's new set of Holmes stories was bound into a volume titled *The Return of Sherlock Holmes.*

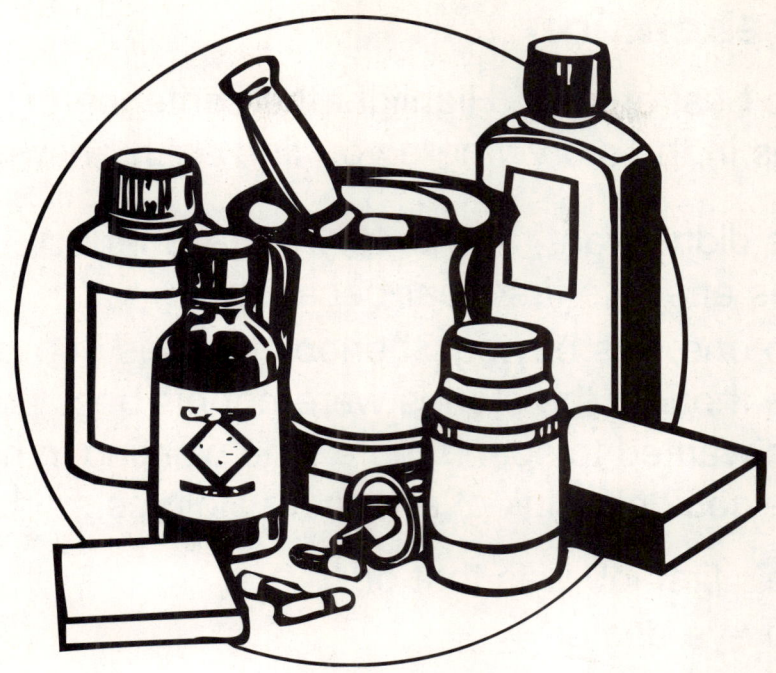

Doyle lived a full life. He liked boxing and motorcar racing. He played football and cricket on a national team. He was a diplomat to Africa. He gave speeches on law reform. He served as a doctor on ships, at war, and in private practice. He fathered five children. He wrote nonfiction works on politics and religion. He wrote fictional tales in all genres. But he would be remembered for just one thing.

Doyle wrote 68 Sherlock Holmes tales. Sales of Holmes stories are second only to those of the Bible worldwide. The character who would not die will not die today. Doyle may have been a fine doctor. He may have written works that were better crafted than his Holmes stories. But to the world, he will forever be the creator of the world's best-known detective.

Comprehension

Circle the best answer. Highlight the sentence or sentences in the story where you find each answer.

1. Doyle didn't want to write any more Sherlock Holmes stories after the first year because . . .
   a. no one was buying Sherlock Holmes books.
   b. he thought the stories were not his best work.
   c. he wanted to spend more time working in his office.
   d. he couldn't think of any more Holmes stories.

2. Doyle's first job was that of . . .
   a. an eye doctor.
   b. a family physician.
   c. a store keeper.
   d. an author.

3. Each of the following statements about Sherlock Holmes stories is true except . . .
   a. Doyle wrote 68 Holmes tales.
   b. only the Bible has sold more copies.
   c. Holmes stories are the best-known works of Doyle.
   d. no one reads Sherlock Holmes stories anymore.

4. Sherlock Holmes's character . . .
   a. resembles one of Doyle's patients.
   b. is named after one of Doyle's friends.
   c. acts like the poet Oliver Wendell Holmes.
   d. resembles one of Doyle's college teachers.

5. The first Sherlock Holmes story . . .
   a. first appeared in a book of short stories.
   b. was named "A Study in Scarlet."
   c. impressed a magazine publisher named Jim News.
   d. was written by Edgar Allan Poe.

The Detective Who Wouldn't Die: Sir Arthur Conan Doyle

6. Sherlock Holmes's assistant . . .
   a. was a college professor named Dr. Watson.
   b. was a medical doctor named Dr. Watson.
   c. was named after a poet.
   d. was killed in a story called "The Final Problem."

7. Sherlock Holmes was brought back to life because . . .
   a. fans, the queen of England, and Doyle's mother asked for more Sherlock Holmes stories.
   b. Doyle missed writing Sherlock Holmes stories.
   c. Doyle needed the money he could make in selling stories.
   d. Hollywood wanted to make a Sherlock Holmes movie.

8. George Newnes . . .
   a. was a banker.
   b. never liked Sherlock Holmes stories.
   c. was a magazine publisher who paid Doyle to write two years' worth of Sherlock Holmes stories.
   d. was one of Doyle's favorite patients.

9. Sir Arthur Conan Doyle . . .
   a. became a rich and famous man.
   b. died poor and unknown.
   c. wrote nothing but Sherlock Holmes stories.
   d. wrote stories because he did not get enough work as a lawyer.

10. Sherlock Holmes stories . . .
    a. are teenage romance novels.
    b. are hard to find today.
    c. are detective tales.
    d. never made Doyle much money.

Name _____ Date _____

Discussion Questions

1. In what ways did Sherlock Holmes's character fit in with Victorian England?

2. Why might Sherlock Holmes stories continue to be popular?

3. Sir Arthur Conan Doyle did many things in his life. Which activities would you find the most interesting?

What Belongs

Listed below are a number of Doyle's life experiences. Also thrown in are a few things Doyle did not do. Cross out all of the activities Doyle did not do in his lifetime.

1. golfed professionally
2. was a doctor on a ship
3. wrote books about religion
4. wrote science fiction books
5. wrote books about dog grooming
6. liked boxing
7. played football
8. wrote mysteries
9. wrote Sherlock Holmes stories
10. lived in America
11. went to Africa
12. worked as a detective

Vocabulary Activities

Match the words on the left with their definitions on the right.

Fill in each blank with a letter to indicate your responses.

_____ **1.** dissuade   a. a partner or friend

_____ **2.** feature   b. demand

_____ **3.** sidekick   c. make to resemble something

_____ **4.** second-rate   d. to give special attention

_____ **5.** insist   e. to soften or yield to an idea

_____ **6.** fashion   f. to talk out of

_____ **7.** relent   g. less than the best

Homework

Read a detective novel of your choice. Write a review of the novel to share with your class.

Extension

In the library or on the Internet, locate a work by Doyle that does not feature Sherlock Holmes. Read it with your class. Do you like Doyle's Holmes stories or his other works better?

Prereading Activities

Looking It Over

1. Read the title of the article that begins on page 36.

2. Leaf through the pages of the article, stopping to look at the pictures.

3. Read the list of vocabulary words.

What Do You Know?

Record here all you know about Abraham Lincoln.

_____

_____

_____

_____

Record here all you know about Thomas Paine and the American Revolution.

_____

_____

_____

_____

Make a Prediction

"Unearthed" means discovered. Reread the title of the article. What do you think the article will be about?

_____

_____

_____

_____

0-7696-3394-3 *Biographies*

American History Unearthed: Abraham Lincoln and Thomas Paine

Vocabulary

surveyed   examined land and figured out boundaries

Example: We and our neighbors surveyed our yards to decide where to put up a fence.

disputed   argued about

Example: A judge decided the disputed land belonged to my grandfather.

legal   within the law

Example: It is legal to turn right on a red light.

elections   the process of voting

Example: Presidential elections are held every four years.

pamphlet   a nonfiction brochure or mini-book

Example: My garden club publishes a pamphlet about fertilizers and pesticides.

skewed   to have bent the truth

Example: Kevin skewed the facts about the number of fish he caught.

infamous   having a bad reputation

Example: The infamous "cookie crook" stole desserts from the school cafeteria.

pauper   a poor, unemployed person

Example: Although he was once a wealthy businessman, my uncle died a pauper.

American History Unearthed: Abraham Lincoln and Thomas Paine

Some school books leave things out. Do you know about the early life of Abraham Lincoln? Do you know about the later life of Thomas Paine? Here's the whole story.

Abraham Lincoln

Lincoln was a great President. He held our country together. He helped end slavery. Tales of his early life hint that he would be an honest and hard worker. They do not suggest he would become a legend.

Lincoln was born in 1809 into hard times. His family suffered failed crops and hunger. They fought land disputes, harsh winters, illness, and death. Lincoln learned to use an axe as soon as he could lift it. He worked on the farm from a young age. So he had little time for school. In fact, he spent less than one year in school. His parents could not read, so Lincoln had to teach himself. And that he did. During all his free time, Lincoln read. He read books in his room. He read books under trees. He read books everywhere.

As a young man, Lincoln worked many jobs. He farmed. He sailed on a riverboat. He was a clerk in a store. When the store went out of business, Lincoln had to look for work again. This time he ran for a seat in the Illinois House of Representatives. He lost. So he opened a new store with a partner. The new store failed, too.

Next, Lincoln held several jobs at once. He had to pay the bills of his failed store. He worked in a post office. He was too nice to be good at that job. He didn't charge poor people for their stamps. He went on long walks to deliver mail to shut-ins in the middle of his shift. When he did, he left the post office door open. He trusted people to pay for their stamps and pick up only their own mail.

In his other jobs, he did well. He surveyed land. He settled land disputes. He filed legal papers for people who couldn't write. He worked on local elections. He worked in a mill and helped build a railroad track.

Then, in 1834, Lincoln ran again for the Illinois House seat. This time he won. He won three more times. He also began practicing law. Lincoln was on a roll. But only for a little while. He won an election that took him to Washington, D.C. There he spoke his mind. Lincoln didn't like the

Mexican War. He said so. People didn't like his views. Lincoln felt like a failure in that job. So he went back to law. But friends changed his mind. In 1860, he ran in one more election. This time he became our sixteenth President. And the rest is history!

Thomas Paine

The life of Thomas Paine was full of ups and downs. It began on the down side. Paine was born in a small town near London. His family was poor. Paine had to leave school and go to work at the age of 13. He didn't like his job. So he went to sea. Paine was a sailor for a few years. Then he returned home. He became a tax collector. After he wrote a pamphlet asking for better pay, he found himself looking for work again. He taught school. He made cigars. He wrote books.

One day, Paine met Ben Franklin. Franklin liked Paine. He helped him get work in the United States. Paine edited the *Pennsylvania Magazine.* He published stories against slavery and for women's rights. He wrote about the needs of old people and children. He also wrote about breaking away from England. His writing *Common Sense* sparked the Revolutionary War.

Thomas Paine was a hero. Every fourth person in the nation owned a copy of *Common Sense.* Paine was asked to help write the Declaration of Independence. When soldiers felt sad, Paine was asked to lift their spirits with words. When words were not enough, he gave money to the war effort. He went to France to ask for their help. America was proud of Paine.

When the war ended, Paine moved to Paris to study bridge design. While there, he supported the French Revolution. But when he said the king should not be killed, he was put in jail. He was on death row. An American ambassador helped him gain his freedom. But Paine was no longer a hero at home.

While in jail, Paine wrote a book. It talked about his beliefs. People did not like his ideas. Newspapers ran lies about Paine's lifestyle. They skewed reports about his beliefs. No one talked about his brave work for the war. When he died in 1809, Paine was an infamous pauper. Only five people attended his funeral. A New York City paper read, "He lived long, did *some* good, and *much* harm." Today we know better.

Comprehension Questions

Circle the best answer. Highlight the sentence or sentences in the story where you find each answer.

1. Thomas Paine and Abraham Lincoln had all of the following in common except . . .
   a. they were both born in the United States.
   b. they were both in politics.
   c. they both had challenging childhoods.
   d. they both learned to work as children.

2. Abraham Lincoln once held a job . . .
   a. as a librarian.
   b. as a teacher.
   c. building railroads.
   d. painting road signs.

3. Abraham Lincoln . . .
   a. never lost an election.
   b. was elected to the Illinois House of Representatives.
   c. was an Indiana congressman.
   d. never worked in Washington, D.C., before he became President.

4. "Unearthed" means . . .
   a. cleaned off.
   b. not on this planet.
   c. discovered.
   d. buried under the ground.

5. Abraham Lincoln was not a very good postmaster because . . .
   a. he was mean to the customers.
   b. he left the post office to deliver mail to shut-ins.
   c. he stole stamps for his own use.
   d. he couldn't read the addresses on envelopes.

6. Thomas Paine was born . . .

   a. in London.

   b. into a rich family.

   c. in the same town as Abraham Lincoln.

   d. after Abraham Lincoln.

7. Thomas Paine was an American hero . . .

   a. at the end of his life.

   b. after he fought in the Revolutionary War.

   c. because his writings sparked the Revolutionary War.

   d. because he supported the North in the Civil War.

8. Thomas Paine wrote . . .

   a. for a magazine called *Pennsylvania Times.*

   b. short stories about the Revolutionary War.

   c. speeches for Abraham Lincoln.

   d. a pamphlet called *Common Sense.*

9. Thomas Paine was saved from death row . . .

   a. when an earthquake destroyed the prison he was in.

   b. by a French ambassador to America.

   c. by an American ambassador to France.

   d. by Benjamin Franklin.

10. Thomas Paine was not popular at the time of his death because . . .

   a. we did not win the Revolutionary War.

   b. people disagreed with his beliefs.

   c. people did not like his wife.

   d. he was mean to his neighbors.

American History Unearthed: Abraham Lincoln and Thomas Paine

Discussion Questions

1. Compare and contrast Abraham Lincoln and Thomas Paine.

2. If Thomas Paine lived today, do you think he would be criticized for having unpopular beliefs? Why, or why not?

3. In your opinion, what was Abraham Lincoln's greatest accomplishment?

Identifying Accomplishments

Below are listed accomplishments of Lincoln and Paine. Write either *Lincoln* or *Paine* before each accomplishment to show who completed it.

_____ 1. Served as a U.S. Congressman

_____ 2. Was born in England

_____ 3. Wrote *Common Sense*

_____ 4. Served as President

_____ 5. Supported the Revolutionary War

_____ 6. Led the North in the Civil War

_____ 7. Wrote about his religious beliefs

_____ 8. Taught himself to read

_____ 9. Was a sailor as a young man

_____ 10. Owned a store

Name _____ Date _____

Vocabulary Activities

Circle the word that better fits in each sentence.

1. Lincoln (surveyed, disputed) land to help find property lines.

2. Tom was (famous, infamous) for his role in ten burglaries.

3. Driving 95 mph on the highway is (legal, illegal).

4. The (disputed, skewed) news stories about Thomas Paine ruined his reputation.

5. Paine's *Common Sense* (elections, pamphlet) inspired colonists to rise up against the king of England.

6. We held (elections, legal) for a new club secretary.

7. The (disputed, skewed) election was finally given to George Bush after much investigation.

8. Thomas Paine died an infamous (pauper, pamphlet).

Homework

Draw a picture of Abraham Lincoln to display in your classroom.

Extension

Locate Thomas Paine's work *The American Crisis.* Memorize and recite to your class the first famous section that begins "These are the times that try men's souls . . . "

Prereading Activities

Looking It Over

1. Read the title of the article that begins on page 46.

2. Leaf through the pages of the article, stopping to look at the pictures.

3. Read the list of vocabulary words.

What Do You Know?

1. Have you ever heard about the Gray Panthers? What do you know about the needs of senior citizens?

   _____

   _____

   _____

2. Record here all you know about unions, strikes, and Mother Jones.

   _____

   _____

   _____

Prediction

What do you think this article is going to be about? What do you think Mother Jones and Maggie Kuhn are likely to have in common? Why do you think so?

_____

_____

_____

_____

Vocabulary

presidential task force

a group of people who advise the President on a certain issue

Example: Maggie Kuhn was a member of a presidential task force on aging.

injustices

laws or circumstances that are not fair

Example: Before the 1960s, minorities experienced many injustices in this country.

epidemic

a disease that affects many people in a given area

Example: A yellow fever epidemic hit the Midwest in the early twentieth century.

union

an organization of workers

Example: The steel workers' union helps improve conditions for steel workers.

dedicated

committed

Example: Teachers are dedicated to the education of children.

strike

refrain from working

Example: The miners voted to strike until they were offered safer working conditions.

45 0-7696-3394-3 *Biographies*

Old Ladies on the Move: Maggie Kuhn and Mother Jones

Maggie Kuhn and Mary Harris Jones made lasting marks on the world. And they both waited until they were old ladies to do it.

Maggie Kuhn

In August 1970, Maggie Kuhn celebrated a birthday. She also mourned the loss of a job. She was 65 years old. Her boss said that was too old. Kuhn had worked since she was in her twenties. She was a teacher. Then she worked at the YMCA for 11 years. Finally she worked in a church for 25 years. Then suddenly she worked nowhere because she was 65.

0-7696-3394-3 *Biographies*

Kuhn was not happy. She loved to work. So did five other friends who had been forced to retire because of their ages. Kuhn and her friends talked about the unfairness of forced retirement. They talked about other concerns that faced older people. They also talked about social concerns that faced people of all ages.

One day Kuhn and her friends met with a group of college students. The old and young talked about their common values and concerns. None of them liked the Vietnam war. None of them liked forced retirement. They all loved America, but they knew some changes could make it better.

Kuhn's gathering of friends grew into a formal group. It had 100 members by 1971. The next year, it adopted a name. A talk show had called Kuhn and her friends the "gray panthers." Many of Kuhn's friends had gray hair. They also fought injustice like panthers. Kuhn took the name for her group.

In 1972, Kuhn gave a speech in front of a lot of people. The public liked what they heard. The Gray Panthers began to grow in power and numbers. Soon they began to work with other groups. They worked with the United Nations. They worked on a presidential task force.

Over the years, the Gray Panthers have stopped injustices. They have made sure the White House holds meetings about aging. They helped stop forced retirements. They have exposed nursing home abuses. They have worked for health care reform. They have fought for peace, housing, and jobs for all people.

At its peak, the Gray Panthers had 60,000 members. Even today, after its founder's death, it still has 50 local networks. And it still strives to perfect living conditions for all.

Mother Jones

Mary Harris Jones was 81 years old when she heard about a coal miners' strike in West Virginia. She had never been inside a coal mine. She had no friends who worked as coal miners. The old dressmaker seemed an unlikely person to lead the strikers. But that is just what she did.

0-7696-3394-3 *Biographies*

In 1912, strikers at the Paint Creek Coal Company in West Virginia welcomed the help of Mother Jones. They were not the first to do so. Jones had understood the troubles of unjustly treated workers since childhood. Her father had been one. He built railroad tracks. His hours were long. His work was dangerous. His paycheck was small. Workers in many other jobs faced the same challenges.

As a young adult, Jones was still worried about workers. Jones married a union leader. He fought for fair and safe working conditions. Jones stayed home and reared four children. Then yellow fever hit Memphis. Rich families moved away. Working-class families had no money to move. So they watched their loved ones fall ill. Jones's husband and children all died in the epidemic.

Jones was alone in the world. She opened a dressmaking shop. She was a good seamstress. Her shop did well. Then tragedy struck again. A fire took Jones's shop and home. Jones looked away from her own troubles. She wanted to help others who worked hard but still had struggles. She dedicated the rest of her life to the union movement. She traveled all over the country. She started new unions. She supported old ones. She addressed strikers. She helped get laws passed to improve the lives of workers. Old Mother Jones owned no more than she could fit in a small suitcase. But she gave a great deal to American workers of the past and present.

49 0-7696-3394-3 *Biographies*

Comprehension

Circle the best answer. Highlight the sentence or sentences in the story where you find each answer.

1. Maggie Kuhn and Mary Harris Jones were both . . .
   a. mothers of ten children.
   b. active into their later years.
   c. child prodigies who began their work at a young age.
   d. famous artists.

2. Which of the following women was forced into retirement?
   a. Maggie Kuhn
   b. Mary Jones
   c. Maggie Kuhn's best friend
   d. Mary Jones's mother

3. Maggie Kuhn . . .
   a. founded the Steel Workers Union.
   b. supported striking coal miners.
   c. founded the Gray Panthers.
   d. worked for unions when her house burned down.

4. All of the tragedies below struck Mother Jones except . . .
   a. her children died of illness.
   b. her husband died of yellow fever.
   c. her home burned down.
   d. her business was lost in a flood.

5. The Gray Panthers . . .
   a. are animals discovered by Maggie Kuhn.
   b. got their name from the youthfulness of their members.
   c. adopted their name from a comment made on a talk show.
   d. are a large labor union.

6. Mother Jones . . .

    a. was born before Maggie Kuhn.

    b. was the founder of the Gray Panthers.

    c. never worked outside of the home.

    d. retired from union work at the age of 65.

7. Mother Jones's father . . .

    a. was a union leader.

    b. was a coal miner.

    c. died of yellow fever.

    d. built railroad tracks.

8. Mother Jones first became concerned about workers . . .

    a. when her husband was killed in a job-related accident.

    b. when her business failed.

    c. when her father endured tough working conditions.

    d. when her mother was harassed on the job.

9. Mother Jones helped the labor movement in all of the following ways except . . .

    a. giving speeches.

    b. starting new unions.

    c. sewing dresses.

    d. supporting old unions.

10. Maggie Kuhn was forced to retire from her job . . .

    a. at a church.

    b. in teaching.

    c. at the YMCA.

    d. at a coal mine.

Old Ladies on the Move: Maggie Kuhn and Mother Jones

Discussion Questions

1. Compare and contrast the lives of Maggie Kuhn and Mary Jones.

2. What hardships faced Mother Jones? How did she handle them?

3. Tell about a retired person whom you know who continues to make contributions to society.

Identifying Main Ideas

Below are three sets of facts from the reading. Determine the main idea presented by each set of facts. Write a main idea sentence in the space provided.

1. _____
   a. Railroad workers worked long hours.
   b. The job of the railroad worker was dangerous.
   c. Railroad workers did not get paid well.

2. _____
   a. Maggie Kuhn liked her job at the church.
   b. She was a good worker.
   c. Maggie Kuhn retired at the age of 65.

3. _____
   a. Mother Jones's husband died.
   b. Mother Jones's children died.
   c. Mother Jones's home and shop burned down.

Name _____ Date _____

Vocabulary Activities

Decide whether each of the following statements is true or false. Indicate your response with a T or F in the spaces provided.

_____ **1.** A *presidential task force* fights wars in other countries.

_____ **2.** Coal miners in Paint Creek suffered *injustices* in 1912.

_____ **3.** Mother Jones supported workers who went on *strike*.

_____ **4.** A *union* member works for animal rights.

_____ **5.** An *epidemic* is another word for a fad or craze.

_____ **6.** Mother Jones was *dedicated* to improving the lives of workers.

Homework

Locate an article in a newspaper about unions or senior citizens. Summarize the article and share your summary in class.

Extension

Research the history of labor unions. Create a timeline of events in the labor union movement.

Old Ladies on the Move: Maggie Kuhn and Mother Jones

Prereading Activities

Looking It Over

1. Read the title of the article that begins on page 56.

2. Leaf through the pages of the article, stopping to look at the pictures.

3. Read the list of vocabulary words.

What Do You Know?

What do you know about the history of America's most popular foods? Record here all you know about any food inventions or popular restaurants.

_____

_____

_____

_____

_____

Make a Prediction

What do you think this article will be about? Why do you think so?

_____

_____

_____

_____

Vocabulary

destined   assigned, or decided in the past

                    Example: Harland Sanders learned to cook at such a young age, he seemed destined to work in food service.

Depression   a time in American history when many people were out of work

                    Example: My grandfather lost his house during the Depression.

streetcar   a vehicle that runs on rails down the street

                    Example: I like to hear the bell ringers who operate the streetcars that run down the rails of San Francisco's Chinatown.

orphanage   a state-run home for children with no parents

                    Example: When his parents died, John moved into an orphanage.

dubbed   named

                    Example: We dubbed our smart teacher "Brainiac."

stock   money invested in a business

                    Example: Colonel Sanders bought the first stock in his new restaurant chain.

Mr. Chocolate Kiss and Colonel Chicken: Milton Hershey and Colonel Sanders

Mr. Chocolate Kiss and Colonel Chicken: Milton Hershey and Colonel Sanders

Two of the great food giants of our times led pretty interesting lives.

Milton Hershey

It did not look like Milton Hershey would do well in life. He only finished fourth grade. He was fired from his first job when he dropped his hat into a printing press. The candy store he opened in 1876 failed.

Hershey did not give up. He liked to make candy. The candy he made was good. He felt destined to run a candy store. So he opened a new one in Denver. It failed. He moved to New York. He opened another candy store. It failed, too. So did the store he opened in Chicago. And the one he tried in New Orleans.

Still, Hershey did not give up on candies. In 1886, he began making caramels. He sold them to candy stores. One day an English candy store placed a large order. Hershey did not have enough ingredients to fill the order. A banker loaned him money. Hershey filled the English order. He filled orders in America, too. He sold so many caramels he paid back his loan early. He also built himself a house and took trips around the world.

On a trip to Germany, Hershey looked at chocolate-making machines. He bought a few to make a chocolate coating for his caramels. His chocolate caramels were a big hit. Soon he stopped making caramels and made only chocolates.

Hershey opened a factory in Pennsylvania. Six hundred people worked for him. Hershey built a fine city for his workers. Hershey, Pennsylvania, had homes, schools, banks, hotels, churches, and a grand park. Hershey's workers were happy. Then the Depression struck. Candy was not selling. The factory didn't need all of its workers. So Hershey gave them construction jobs. They built a sports arena and community center in Hershey. They built a senior hall and new offices. They made their town an even better place to live.

57 0-7696-3394-3 *Biographies*

Hershey made life nice for more than just his workers. He built an orphanage for boys. He funded museums. He opened a junior college. He built a library and a hospital. Hershey is famous for his chocolate, but he gave much more to this world!

Harland Sanders

In 1896, six-year-old Harland Sanders's father died. The child's mother had to go to work. Harland had to learn to cook. He was the oldest of three children. He had to watch his siblings and cook them meals. By the time he was seven, Sanders was a great cook. Still, he did not cook for a living for another 34 years.

From age 10 until 15, Sanders worked on a farm. Then he drove a streetcar for a year. Next, Sanders became a private in the army. From age 16 to 24, Sanders was a busy man. He served as a justice of the peace. He sold insurance. He ran a ferry. Then he became the owner of a gas station.

Sanders's gas station in Corbin, Kentucky, was a popular stop. That's because you could get gas at a good rate there. You could also buy a fine meal in the back room. The food Sanders served was good. People started stopping at the gas station just to eat Sanders's chicken.

So Sanders changed jobs again. He bought a motel and café across the street from the gas station. The public was pleased. The governor of his state dubbed Sanders "Kentucky Colonel." His café won awards.

Then a highway was built. Travelers were routed right past Corbin. No one came to buy his chicken anymore, so Sanders took his chicken to them. He drove all over the country. He carried packets of spice mixes in his back seat. In each town, he offered to fry chicken for restaurant owners. They liked what they tasted. They ordered more packets of the spices. Sanders called home. His wife mixed the spices and shipped them by train to fill her husband's orders.

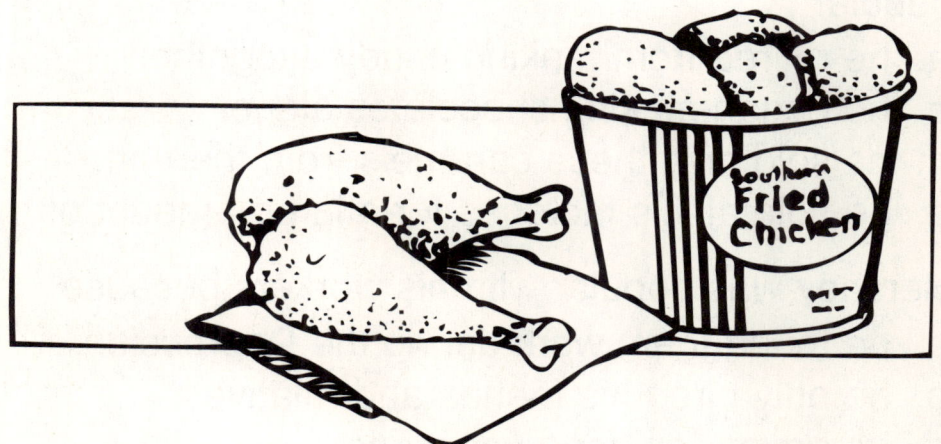

By 1964, hundreds of places served Kentucky Fried Chicken®. So the restaurants were made into a chain. Sanders bought the first 100 shares of stock in the chain. The six-year-old cook had grown into a million-dollar chef.

 0-7696-3394-3 *Biographies*

Comprehension

Circle the best answer. Highlight the sentence or sentences in the story where you find each answer.

1. It did not appear Milton Hershey would do well in life because . . .
   a. he failed college.
   b. he didn't make very good candy.
   c. the candy stores he opened all failed.
   d. he spent his first ten adult years in prison.

2. Hershey once ran a candy store in . . .
   a. New York.
   b. Chicago.
   c. Denver.
   d. all of the above as well as New Orleans.

3. Hershey's chocolate-covered caramels became so popular . . .
   a. he retired from making candy altogether.
   b. he began making chocolates alone.
   c. he sold only these caramels from then on.
   d. he sold all his factories for a large amount of money.

4. Hershey was popular with his workers because . . .
   a. he found them work during the Depression.
   b. he only hired his friends and relatives.
   c. he gave them long vacations.
   d. he never made them work.

5. The town of Hershey . . .
   a. is located in New York State.
   b. has museums and a junior college.
   c. is where Milton Hershey grew up.
   d. became a ghost town during the Depression.

6. When Harland Sanders was six years old, . . .

    a. his house burned down.

    b. his grandma died.

    c. his father died.

    d. his youngest brother was born.

7. Sanders worked on a farm until he was . . .

    a. 15 years old.

    b. 20 years old.

    c. 12 years old.

    d. 65 years old.

8. Which statement is true about Harland Sanders?

    a. The governor of Texas dubbed him a colonel.

    b. He worked in food service his entire life.

    c. He sold packets of chicken coating out of his car.

    d. He never did any of his own cooking.

9. The most popular item at Sanders's gas station became . . .

    a. gas.

    b. auto parts.

    c. chicken.

    d. pies.

10. Which statement about Harland Sanders is not true?

    a. Sanders was in the U.S. army.

    b. Sanders was a married man.

    c. Sanders learned to cook at age 10.

    d. Sanders once owned a motel and café.

Mr. Chocolate Kiss and Colonel Chicken: Milton Hershey and Colonel Sanders

Discussion Questions

1. Compare and contrast the lives of Colonel Sanders and Milton Hershey.

2. Hershey felt destined to work with candy. What do you think you will do with your adult life? Do you feel destined to your dream?

3. In what ways did Sanders and Hershey turn challenges into opportunities?

Adjective or Adverb

Adjectives tell more about nouns. Adverbs tell more about adjectives, verbs, or other adverbs. Decide whether each of the following words in italics is an adjective or an adverb. Indicate your responses in the spaces provided.

_____ 1. Restaurant owners ordered *more* packets of spices.

_____ 2. You could also buy a *fine* meal in the back room.

_____ 3. Hershey is famous for his chocolate, but he gave *much* more to this world.

_____ 4. They made their town an *even* better place to live.

Vocabulary Activities

Circle the letter of the unrelated word in each set.

1. a. destined   b. decided   c. assigned   d. free

2. a. Depression   b. recession   c. economy   d. America

3. a. streetcar   b. rocket   c. trolley   d. train

4. a. orphanage   b. shelter   c. bus station   d. home

5. a. dubbed   b. named   c. scrubbed   d. called

6. a. stock   b. bonds   c. shares   d. clothing

Homework

Write an imaginary story about the history of your favorite food. Who invented it? How did he or she stumble upon the invention?

Extension

Locate the addresses of Kentucky Fried Chicken® headquarters and Hershey® Chocolates headquarters. Write both places a letter commenting about their products.

Prereading Activities

Looking It Over

1. Read the title of the article that begins on page 66.

2. Leaf through the pages of the article, stopping to look at the pictures.

3. Read the list of vocabulary words.

What Do You Know?

1. Who is Babe Ruth? What stories do you know about his life or career?

_____

_____

_____

_____

2. Have you ever heard of Babe Didrikson? If so, what do you know about her?

_____

_____

_____

_____

Make a Prediction

What might this story be about? Why do you think so?

_____

_____

_____

Vocabulary

rigged   influenced for dishonest purposes, involved cheating

Example: When Lisa won a third time in a row, I began to think the game was rigged.

coaxed   urged or persuaded

Example: Karen was coaxed into the cold pool by her pleading brother.

prone   having a tendency

Example: Because of her all-around skills, Babe Didrikson was prone to win everything she played.

surly   cross and disrespectful

Example: When Babe Ruth acted surly with his parents, he was sent to St. Mary's School.

exhibition   show or display

Example: Although it is not league play, this exhibition game is quite fun to watch.

succumbed   gave in to, surrendered

Example: Babe Didrikson succumbed to cancer at an early age.

Babes at Bat: Babe Ruth and Babe Didrikson

Babe Ruth and Babe Didrikson were both athletes. It didn't seem likely either would make the big leagues. Didrikson had too few chances to play. Ruth's trouble-making seemed likely to keep him out of the game.

Babe Ruth

The year was 1920. Baseball was at an all-time low. The last World Series had been rigged. People had lost respect for baseball. Then a great Boston Red Sox pitcher was sold to the New York Yankees. His smile and his swing coaxed fans back to the game.

In 1902, no one could have guessed George Ruth would one day be a baseball legend. That year the seven-year-old was sent to live with priests. Ruth was already chewing tobacco, cussing, and acting tough. His parents hoped the church could turn the boy around.

66 0-7696-3394-3 *Biographies*

Brother Matthias did more than that. He taught Ruth to read, write, and play baseball. Ruth learned to play ball so well, in fact, that the manager of a pro team came to watch him play. Jack Dunn liked what he saw. He asked Ruth to play with the Baltimore Orioles. At 19, Ruth did. The coach had found himself a new "babe."

Next, Ruth played for the Boston Red Sox. With them, he was one of the best pitchers of his day. But more amazing was his swing. He was so prone to hit home runs, he began to play outfield. Ruth's batting talent was too valuable. He had to play a position that allowed the most batting time.

The Red Sox coach made a good choice. In 1919, Ruth set a new homerun record at 29. The next year, he joined the Yankees. While with them, he broke his own record. This time he hit 60 home runs in a season. He became known as the "Sultan of Swat." He led his league in home runs during 12 seasons. In all, he belted 714 home runs.

Babe Ruth became a star. He awed his fans on and off the field. Whatever made him a wild child stayed with him as an adult. Ruth smashed up fast cars. He argued loudly with his coach. He ate and drank too much. He stayed up all night on game nights. He also showered his fans with smiles, waves, and autographs.

Babe Ruth changed the face of baseball. Hitters no longer played it safe. Fans wanted to see hard hitters. They wanted to see the Babe. The cussing, surly child had grown into a big hit.

The Other Babe

Mildred Didrikson was born in 1914. "Babe" grew up in a large, active family. As a child, she ran races. She roller-skated. She played ball in a gym her dad built in the back yard. She earned her nickname in grade school when she hit five home runs in one game. By high school, she excelled in volleyball, tennis, baseball, basketball, and swimming.

There were no major league ball teams for women in the 1930s. Still, Babe found ways to play sports. At 18, she led her work's basketball team to a national championship. That year she also broke four world records in amateur track events.

In 1932, Babe qualified for five Olympic events. Women could play in only three then. She won the javelin gold. She won the 80-meter hurdles gold. She broke a high jump record. The judges didn't like her style. They gave her the silver.

Next, Babe toured the nation. She played in exhibition games. She played basketball, baseball, and tennis. She bowled and shot rifles. She swam and dove. She roller-skated and played volleyball. She struck out Lou Gehrig in spring training. She challenged Babe Ruth to a game of golf.

In 1933, Babe settled into playing just golf. She hit thousands of balls a day. Her hands blistered and bled. She taught herself to hit 250 yards on a regular basis. Then she won 80 amateur tournaments. In 1945, she started the Ladies Pro. Golf Association. She went on to win 31 LPGA events.

In 1950, Babe was named the "Greatest Athlete of the First Half of the Century." Then she succumbed to cancer. She was only 42 years old. Her short life had been packed full. Some sports scholars think Babe was the best all-around athlete—male or female—the world has ever seen.

0-7696-3394-3 *Biographies*

Comprehension

Circle the best answer. Highlight the sentence or sentences in the story where you find each answer.

1. Babe Didrikson earned her nickname . . .
   a. because she was a big Babe Ruth fan.
   b. when she hit five home runs in a single game.
   c. during her college days.
   d. because she acted like a baby.

2. Babe Ruth was sent to live with priests . . .
   a. when he was 10 years old.
   b. because his parents died.
   c. because he was an unruly child.
   d. to get a good education.

3. Babe Ruth . . .
   a. did not learn to play baseball until he was 30.
   b. began playing major league ball when he was 25.
   c. learned to play ball from Brother Matthias.
   d. did not like his fans.

4. Babe Ruth started his career as a . . .
   a. pitcher.
   b. catcher.
   c. outfielder.
   d. first baseman.

5. Babe Ruth became . . .
   a. the "Greatest Athlete of the First Half of the Twentieth Century."
   b. a triple medal Olympiad.
   c. the single-season home run record holder in 1919 and 1927.
   d. a respected baseball coach.

6. In her last years, Babe Didrikson did most of her playing . . .
   a. on the tennis court.
   b. on the basketball court.
   c. in exhibition games.
   d. on the golf course.

7. Babe Didrikson did not play . . .
   a. table tennis and hockey.
   b. volleyball and basketball.
   c. golf and tennis.
   d. track and field events.

8. Babe Didrikson died . . .
   a. at the age of 93.
   b. of cancer at the age of 42.
   c. before the creation of the LPGA.
   d. as the result of a sports injury.

9. How did the women's sports world differ in the 1930s?
   a. Women could compete in only five Olympic events.
   b. There were no women's professional leagues.
   c. Women could play golf only professionally.
   d. Women and men shared a baseball league.

10. Which statement about Babe Didrikson's childhood is true?
    a. Her father built a gym in her back yard.
    b. She was an only child.
    c. She was a sickly, inactive child.
    d. She did not enjoy playing sports.

Name _____ Date _____

Discussion Questions

1. Compare and contrast the careers of Babe Ruth and Babe Didrikson.

2. How did the two Babes get their nicknames? Do you or your friends have nicknames? Share the stories behind them.

3. Who is your favorite professional athlete? What was your own most memorable sports moment?

Compare and Contrast

How are each of the following pairs of sports alike? How are they different? Write your responses in the spaces provided.

	How they are alike	How they are different
basketball and volleyball		
tennis and golf		
baseball and track		
hockey and soccer		

Vocabulary Activities

Antonyms are opposites. Synonyms are similar. Decide whether the following pairs of words are antonyms or synonyms. Circle your responses.

1. rigged: fixed   antonym   synonym

2. coaxed: discouraged   antonym   synonym

3. prone: inclined   antonym   synonym

4. surly: kind   antonym   synonym

5. exhibition: show   antonym   synonym

6. succumb: surrender   antonym   synonym

Homework

Write a one-page account of your favorite moment in sports history.

Extension

Research and report on an athlete of your choice.

The Greatest Showman on Earth: P. T. Barnum

Prereading Activities

Looking It Over

1. Read the title of the article that begins on page 76.

2. Leaf through the pages of the article, stopping to look at the pictures.

3. Read the list of vocabulary words.

What Do You Know?

Have you ever been to the circus? What did you see there?

_____

_____

Have you ever heard the name P. T. Barnum? What do you know about him?

_____

_____

Make a Prediction

P. T. Barnum was called the "Greatest Showman on Earth." What do you think he contributed to the circus world? What other contributions might he have made to the world of entertainment?

_____

_____

_____

_____

Vocabulary

blunt   to the point, lacking subtlety

Example: Barnam's blunt comments offended some people.

autopsy   a process used to determine the cause of death and condition of a dead body

Example: An autopsy indicated Joice Heth was 80 years old at the time of her death.

touted   praised, advertised

Example: Kevin touted his chess-playing skills so loudly no one would dare play him.

basked   enjoyed the warmth

Example: Barnum basked in his success at the end of his life.

petrified   turned into a stony substance

Example: Petrified wood is hard as rock.

lecture   an informative talk

Example: Kaley took notes as her teacher gave a lecture on P. T. Barnum.

The Greatest Showman on Earth: P. T. Barnum

When P. T. Barnum was 16, his father died. Barnum was now the man of the family. He tried his hand at many trades. He sold hats. He ran a boarding house. He ran a store. He sold lottery tickets. Then he put out a newspaper. One of the stories it ran was so blunt and unkind, it landed Barnum in jail. When he got out, a parade of supporters greeted him. The experience made Barnum realize two things: the media can stir emotions, and Barnum was in the wrong job.

Barnum sold his newspaper. He sank all of his money into Joice Heth. Heth said she was a 161-year-old former slave. She claimed to be George Washington's former nanny. Barnum toured the country with Heth. In every town a big band marched along Main Street. Then Barnum and Heth stopped. They charged people to listen to Heth's tales. Then Heth died. An autopsy showed she was only 80 years old.

Next, Barnum touted a former slave who juggled and sang. In each town, someone challenged Barium's juggler to a match. In each town, the challenger lost. That's because Barnum paid him to lose.

Barnum showcased his juggler in towns. He also had him perform on a paddleboat. Then Barnum bought a museum. Other performers joined his juggler there. There was a singing dwarf. There were Siamese twins. There was a tattooed man and a bearded lady. Beside the live acts were other oddities. The museum housed a petrified giant. It also displayed a mermaid with a fish's tail and a monkey's head.

Barnum's museum made a lot of money. Barnum used the money to build a new town. One of the businesses he supported went bankrupt. Barnum went down with it. He had to sell his museum. His house burned down.

So Barnum hit the road again. This time he set out on a lecture tour. He told people about his life. He gave them tips on making money.

Barnum's lecture tour made him money. He bought his museum back. It burned to the ground. He built it up again. It burned down again. Sixty-year-old Barnum retired. Until . . .

A friend asked Barnum to take his show on the road again. Barnum was now in the circus world. His show soon merged with Bailey's show. The Barnum and Bailey circus boasted a train, 500 workers, and 200 horses. It brought in over one million dollars a year. Barnum basked in his success. He had become the "Father of Modern Advertising" and the inventor of the paying crowd.

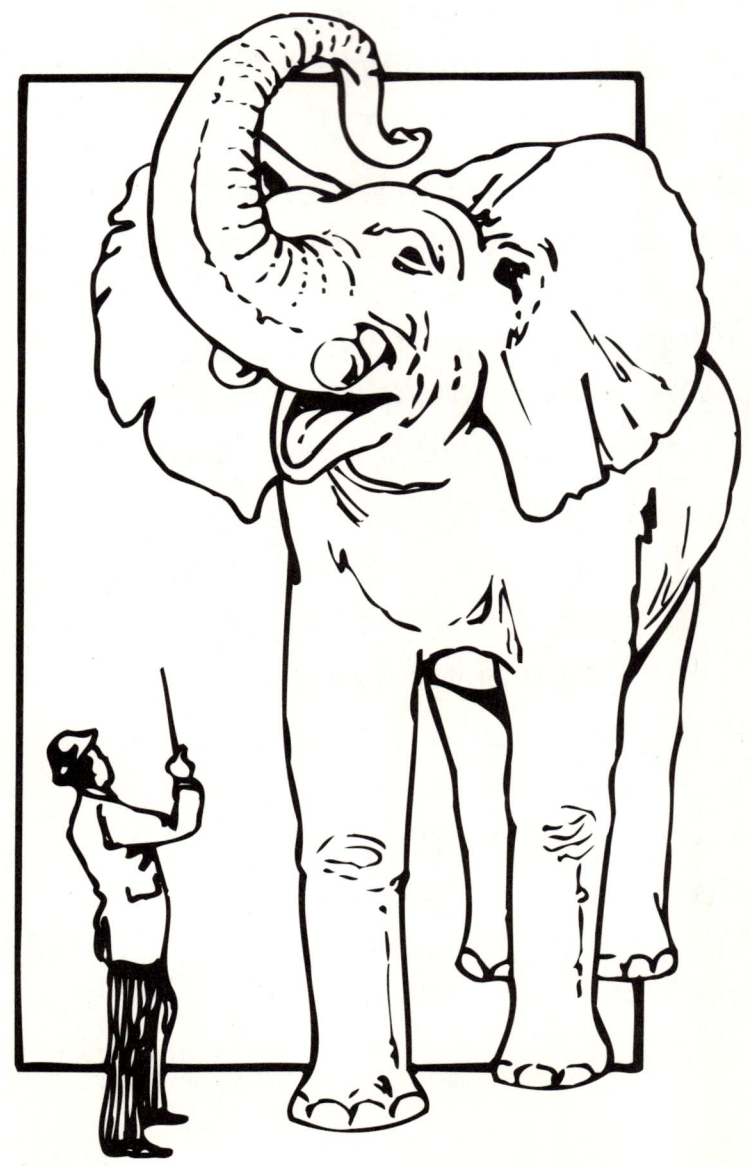

0-7696-3394-3 *Biographies*

Comprehension

Circle the best answer. Highlight the sentence or sentences in the story where you find each answer.

1. Joice Heth was . . .
   a. once the nanny of George Washington.
   b. lying about her age and her past life.
   c. 61 years old at the time of her death.
   d. P. T. Barnum's wife.

2. P. T. Barnum was a great . . .
   a. newspaper editor.
   b. store operator.
   c. hat salesman.
   d. showman.

3. P. T. Barnum once spent six months in jail for . . .
   a. writing unkind things in his newspaper.
   b. lying about the age of Joice Heth.
   c. charging too much for his shows.
   d. refusing to pay his taxes.

4. P. T. Barnum's museum . . .
   a. housed dinosaur bones.
   b. burned down three times.
   c. displayed strange live acts and other curiosities.
   d. is still around today.

5. The people who challenged Barnum's juggler never won because . . .
   a. Barnum's juggler was a professional.
   b. Barnum paid the challengers to lose.
   c. Barnum never let audience members juggle.
   d. Barnum only let children challenge his juggler.

Answer Key

A Life of Adventures: Miguel de Cervantes Saavedra • Rags to Riches: Hans Christian Andersen and Alice Walker • The Detective Who Wouldn't Die: Sir Arthur Conan Doyle
American History Unearthed: Abraham Lincoln and Thomas Paine • Old Ladies on the Move: Maggie Kuhn and Mother Jones
Mr. Chocolate Kiss and Colonel Chicken: Milton Hershey and Colonel Sanders • Babes at Bat: Babe Ruth and Babe Didrikson • The Greatest Showman on Earth: P. T. Barnum

A Life of Adventures.......................pages 10–13

1. c	6. a
2. a	7. d
3. d	8. c
4. b	9. a
5. b	10. c

Sequencing: 6, 2, 1, 4, 3, 5

1. malarial	4. ransom
2. antics	5. scholars
3. scrounge	6. armada

Rags to Riches................................pages 20–23

1. d	6. b
2. b	7. c
3. a	8. c
4. d	9. d
5. c	10. a

The Detective Who Wouldn't Die...pages 30–33

1. b	6. b
2. a	7. a
3. d	8. c
4. d	9. a
5. b	10. c

What Belongs: cross out 1, 5, 10, 12

1. f	5. b
2. d	6. c
3. a	7. e
4. g	

American History Unearthed ....pages 40–43

1. a	6. a
2. c	7. c
3. b	8. d
4. c	9. c
5. b	10. b

Identifying Accomplishments: Lincoln, 1, 4, 6, 8, 10; Paine, 2, 3, 5, 7, 9

1. surveyed	5. pamphlet
2. infamous	6. elections
3. illegal	7. disputed
4. skewed	8. pauper

Old Ladies on the Move ....pages 50–53

1. b	6. a
2. a	7. d
3. c	8. c
4. d	9. c
5. c	10. a

Identifying Main Ideas: 1. Railroad workers suffered harsh working conditions.
2. Maggie Kuhn was forced into retirement.
3. Mother Jones suffered personal hardships.

1. F	4. F
2. T	5. F
3. T	6. T

Mr. Chocolate Kiss and
Colonel Chicken............................pages 60–63

1. c	6. c
2. d	7. a
3. b	8. c
4. a	9. c
5. b	10. c

Adjective or Adverb: 1. adjective 2. adjective 3. adverb 4. adverb

1. d	4. c
2. d	5. c
3. b	6. d

Babes at Bat ....pages 70–73

1. b	6. d
2. c	7. a
3. c	8. b
4. a	9. b
5. c	10. a

1. synonym
2. antonym
3. synonym
4. antonym
5. synonym
6. synonym

The Greatest Showman on Earth ....page 79

1. b
2. d
3. a
4. c
5. b

0-7696-3394-3 *Biographies*